MW00715617

An Ant Colony

An Ant Colony

by Heiderose and Andreas Fischer-Nagel

A Carolrhoda Nature Watch Book

Carolrhoda Books, Inc./Minneapolis

Thanks to Dr. Basil Furgala and Professor Ralph W. Holzenthal, Department of Entomology, University of Minnesota, for their assistance with this book

For our children, Tamarica and Cosmea Désirée

Photograph on page 8 courtesy of Edward S. Ross, California Academy of Sciences

This edition first published 1989 by Carolrhoda Books, Inc.
Original edition copyright © 1985 by Kinderbuchverlag KBV Luzern AG, Lucerne, Switzerland, under the title DER AMEISENSTAAT.
Translated from the German by Gerd Kirchner.
Adapted by Carolrhoda Books, Inc.
All additional material supplied for this edition ©1989 by Carolrhoda Books, Inc.

LIBRARY OF CONGRESS CATALOGING-IN-PUBLICATION DATA

Fischer-Nagel, Heiderose.
[Ameisenstaat. English]
An ant colony / by Heiderose and Andreas Fischer-Nagel ; [translated from the German by Gerd Kirchner].
p. cm.
Translation of: Der Ameisenstaat.
"A Carolrhoda nature watch book."
Includes index.
Summary: Describes the life cycle and community life of ants.
ISBN 0-87614-333-8 (lib. bdg.)
1. Ants—Behavior—Juvenile literature. 2. Insect societies—Juvenile literature. 3. Insects—Behavior—Juvenile literature. [1. Ants.] I. Fischer-Nagel, Andreas. II. Title.
QL568.F7F4613 1989
595.79'604248—dc19 88-31564
CIP
AC

Manufactured in the United States of America

2 3 4 5 6 7 8 9 10 98 97 96 95 94 93 92 91 90

Many of us have watched ants scurrying about on trees, in the grass, on the sidewalk, and even in our homes. We've marveled at their ability to carry insects and twigs much larger and heavier than they are themselves. Although ants appear to be confused and disoriented, they follow specific paths and perform specific tasks.

Ants can be found on every continent in the world. They can adapt to any temperature except the extreme cold of the polar regions. There are about 8,000 species of ants known to scientists and more are being discovered all the time.

Different ant species have adapted to different ways of living. Slave maker ants raid other ants' nests, take their young, and raise them to be slaves. Harvester ants gather seeds and store them in their nests. Army ants hunt other insects, and dairying ants "milk" certain insects that produce a tasty liquid. In this book, we will observe the red ant. There are many species of red ant, but for our purposes, whenever we refer to red ants, we're referring to the species, *Formica polyctena*. We will learn much about general ant behavior by observing the red ant's way of living.

Like bees and termites, ants are social insects. They live in organized communities, called **colonies**, that may consist of hundreds to millions of members. Every member of a colony has a certain task to perform. Each ant depends on all the other ants to do their own specific jobs. Without its colony, an individual ant could not survive.

Three types of ants are found in every colony: queens, males, and workers. The **queens** are much larger than the other ants. Their primary task is to lay eggs for the colony. Some colonies have only one queen. But red ants may have up to 5,000 queens in a single colony. Red ant queens are one of the longest-lived insects, and during their life span of 20 to 25 years, they lay tens of thousands of eggs. Queens have wings until they mate.

Male ants also have wings. Their only duty is to mate with the queens. Males are the shortest-lived ants in a colony. They hatch in the spring, mate in the summer, and die as soon as their job is done.

Most of the ants in a colony are females called **workers**. Living for five or six years, red ant workers lead busy lives. They are responsible for building, repairing, and defending the nest, and for caring for the queens and the **brood**—the eggs, larvae, and pupae. Workers also gather food and are responsible for feeding all the members of the colony. Some workers perform the same task their entire lives, others are constantly changing tasks.

Ants spend the winter protected by their nest, or anthill. Some ant species build nests from tree leaves, some build nests inside plants, some tunnel underground, and some have no permanent nests at all. The size of the anthill depends on whether the area is sunny or shady. The more surface area an anthill has, the more of the sun's warming rays it can absorb. So anthills built in the shade are taller than those built in the sun. Warmth and humidity are carefully regulated in a nest so that the brood will develop properly.

Not only the sun, but also the ants themselves can supply heat for their colony. In spring, as soon as the snow has melted away from the anthill, the first ants emerge to bask in the sunlight. If the weather is warm enough, nearly the whole colony comes out, including the queens, to absorb the warmth. At other times of the year, only the workers bask in the sun's heat until their bodies warm to a temperature of 86° to 104° F (30° to 40° C). Then, when they go into the anthill, the heat is released, warming the nest. Another way ants regulate the nest's temperature is to protect the colony from cold air by blocking the entrances to the nest with their bodies.

On a warm summer day, all the young queens and all the males leave their nests to embark on their mating flight. The picture to the upper left shows a young ant queen shortly before the mating flight. She still has her wings. The mating flight is the only time that male and queen ants fly. A queen may mate with one male or many males. After mating, queens have a lifetime supply of **sperm**, or male reproductive cells, stored in their bodies.

When the mating flight is over, the ants drop to the ground. The males die, and the queens search for a nesting site. Some of these young red ant queens will return to their home nest, some will find another established nest to inhabit, but some will look for a tree stump in the forest to start their own colony. The now useless wings either drop off or are bitten off by the workers. Queens that still have wings in the late summer have not mated. A red ant queen seeking to establish a new colony will find a tree stump, bore a hole into it, and cover the hole with twigs, pine needles, and dry leaves. Then she lays her **eggs**. Since the queen must stay inside the nest to tend her eggs, she cannot go out to look for food. She survives by absorbing nutrition from the wing muscles she no longer uses and by eating some of her own eggs.

The first ants that hatch are workers. They quickly venture outside to gather food so they can feed the queen. From this point on, workers take care of the queen by licking and feeding her. Now her only job is to lay eggs.

After the queen has been tended to, the workers begin to enlarge the anthill by digging tunnels and chambers in the soil. These chambers are where the eggs and the little **larvae** are cared for. Larvae are ants in their second stage of development. Throughout this stage, larvae increase in size. The big larvae and **pupae**, or ants in their third stage of development, live above the ground, on and around the stump, where the workers have built chambers made of coarse plant debris and twigs. As the anthill is built up, finer materials are used.

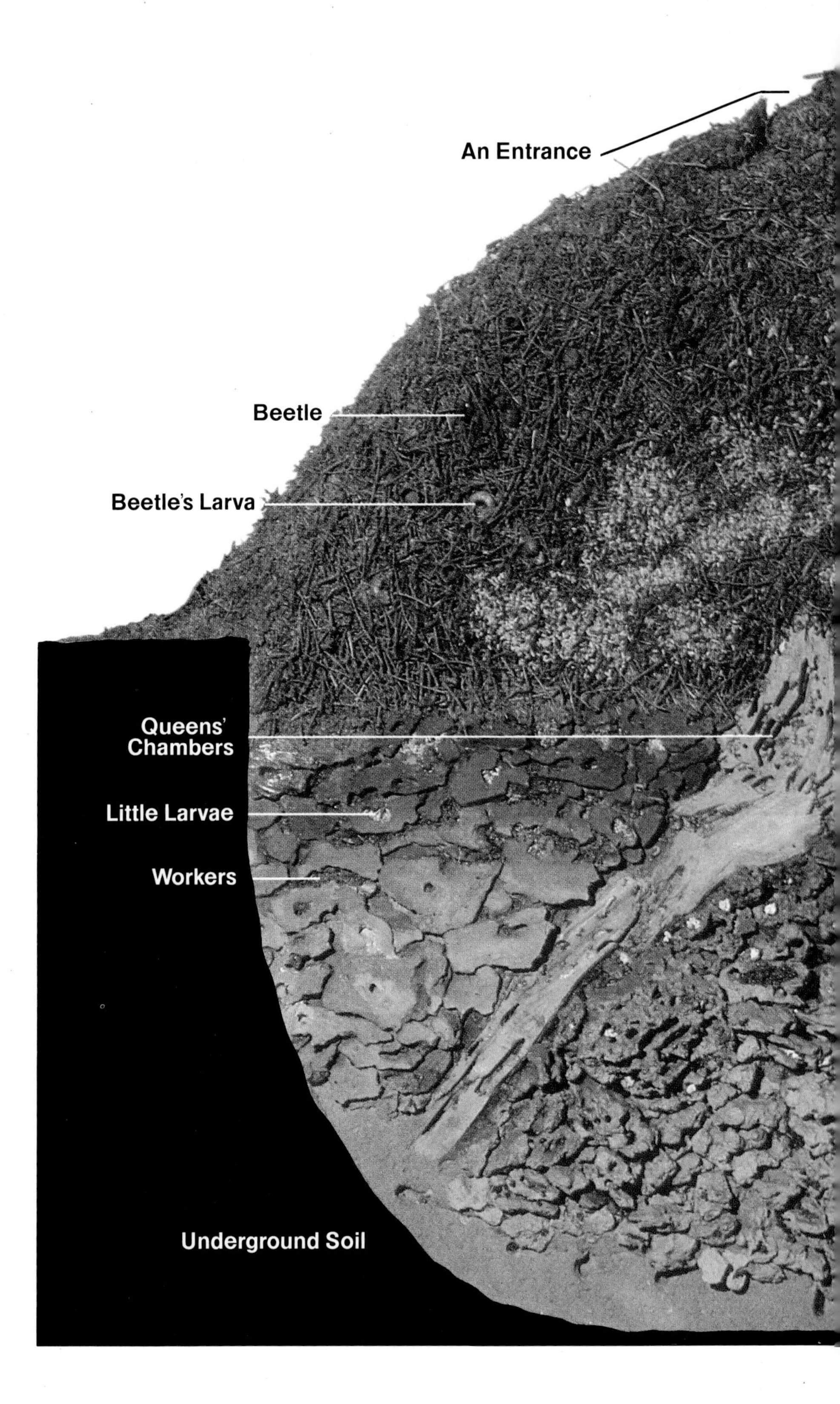

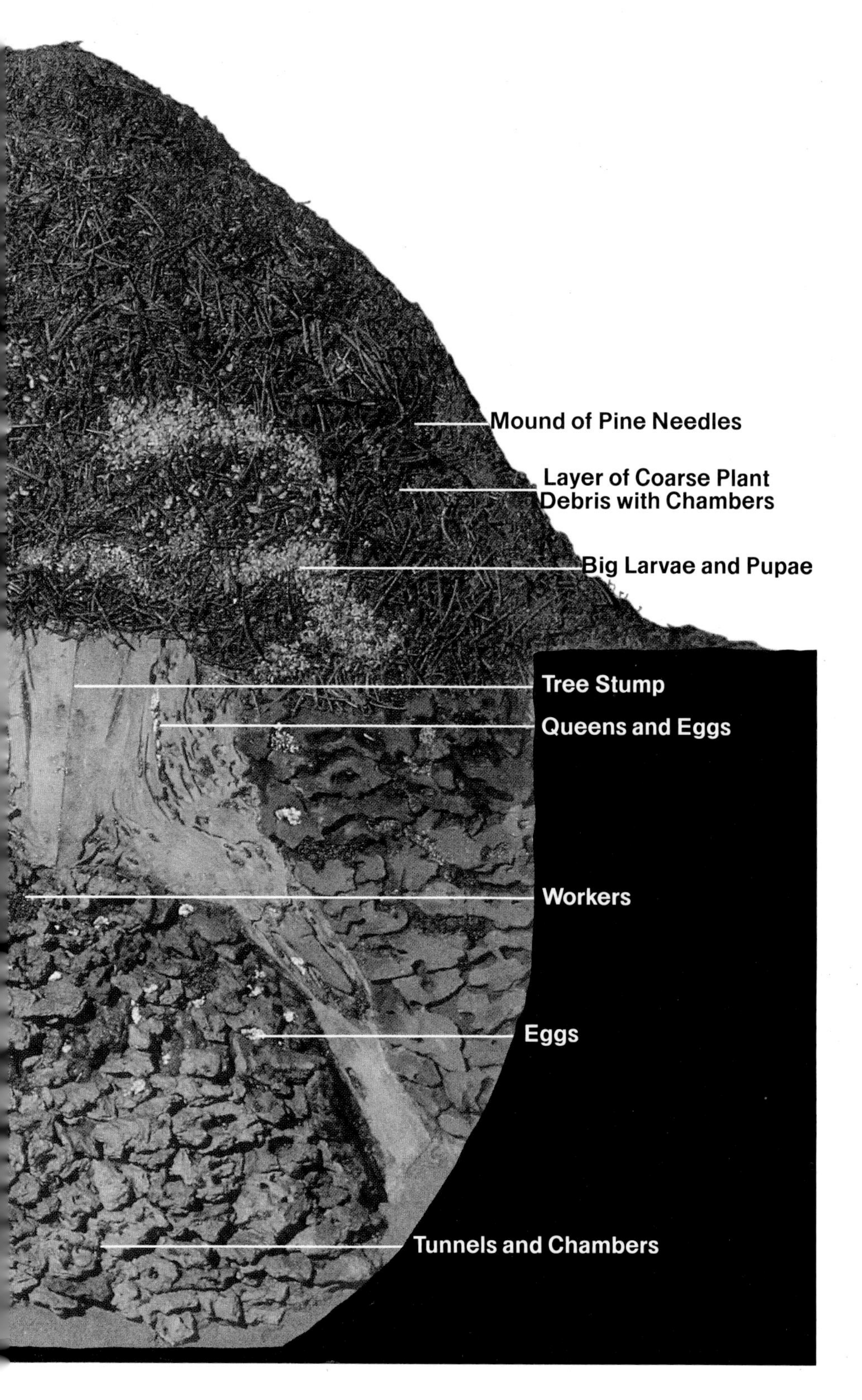

The outermost protective layer is made of pine needles. Insects, such as some kinds of beetles, might make their homes in this layer. The amazingly skillful construction of the entrances into an anthill allows for even the lowermost chambers, deep in the ground, to be regularly supplied with fresh air.

Red ant queens usually live in these lowermost chambers inside a tree stump, where they also lay their eggs. These queens each lay about 10 eggs a day. The stored sperm enter the eggs as they are being laid. Since there are up to 5,000 queens in a red ant colony, there may be thousands of eggs in the nest at one time. The moment the tiny eggs are laid, they are carried to special chambers, called nurseries, by worker ants.

The many chambers in an anthill differ in temperature and humidity. Those used as nurseries must have a temperature of 77° F (25° C) for the eggs to develop properly. Nurses, worker ants who care for the brood, move the eggs from one chamber to another as temperatures fluctuate. They also keep the eggs clean and moist by constantly licking them. Ant saliva is sticky, so when the eggs are licked, they stick together in clumps, making them easier to be carried.

The first stage of development is complete after 14 days, when the tiny larvae hatch. Without legs or eyes, the larvae do not look like adult ants at all. The larvae depend on the nurses to feed and clean them. While they are small, the larvae are stuck together with saliva so they can easily be carried to different chambers where the temperature is about 82° F (28° C) and the humidity is very high. The food for the larvae consists of a special liquid meat paste the nurses receive from food-gathering ants.

Within the next 8 to 20 days, the larvae grow quickly. When their skin gets too tight for them, they literally burst at the seams and crawl out. After the larvae have shed their skin four or five times, they begin the next transformation—they **pupate**, or become pupae.

The larvae produce a liquid that slowly solidifies when it comes into contact with the air. This is spun into a protective coating, or **cocoon**. The cocoon is white and looks much like the eggs, but bigger. For no known reason, there are always a certain number of larvae that do not spin cocoons. After these larvae have shed their skin for the last time, they also pupate. These pupae lie helplessly in the nest (shown in the lower left picture) without a protective coating. Their legs and antennae are tightly pressed to their bodies, and they are colorless. The pupae in the cocoons have the same appearance.

During this stage, all the pupae are taken by the workers to dry places that are about 86° F (30° C). These places are usually right below the surface of the anthill, between little sticks and twigs, where they may be seen from the outside. Ant pupae are often mistaken for ant eggs. Eggs, however, are so tiny they can hardly be seen, and eggs can only be found very deep within the nest.

After two to three weeks in the cocoon, the transformation is complete. The young ant is ready to hatch, and it gnaws a hole from within its cocoon (top picture). As soon as the nurses notice this, they help by widening the hole (middle picture). A clumsy young ant hatches out of the cocoon (bottom picture). For the first few days, its body is still soft and vulnerable, and its adult coloration hasn't yet developed—its **thorax**, or chest, is light brown, its legs are pale, its head and abdomen are gray. If the anthill is threatened, nurses will grab the young ants and carry them to safety.

This is a life-size picture of a red ant. You have to look very closely to see the details of its body. Compared to some ants, red ants are quite big. Yet they measure only 1/3 inch (8 mm)! The enlargement to the right makes it easier for us to see all the details. Ants, like other insects, have a hard covering called an **exoskeleton**, which protects the internal organs. Their bodies are divided into three parts: the head, the thorax, and the abdomen.

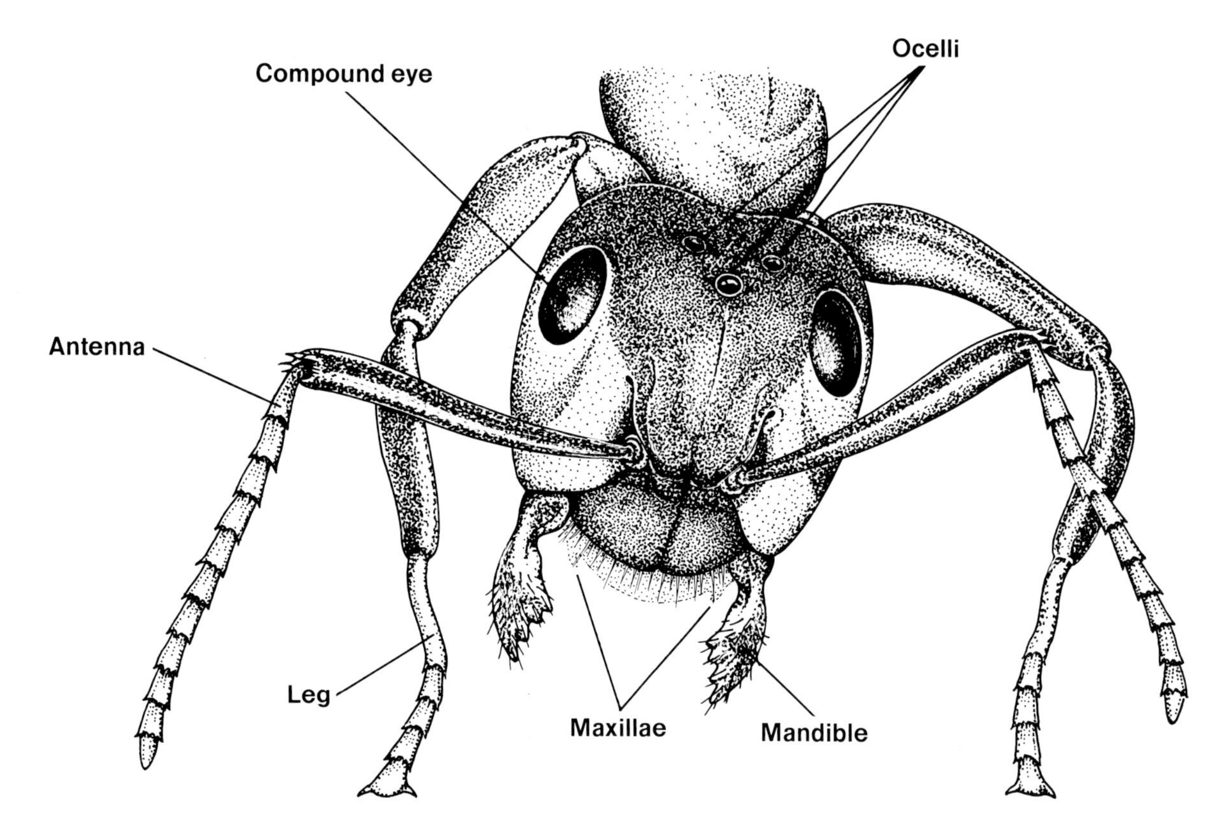

On the head are antennae, eyes, and mouthparts. **Antennae** are tiny sense organs that allow ants to touch, taste, smell, and feel vibrations. Ants also use their antennae to communicate with each other.

Eyes are another type of sense organ. All ants have two **compound eyes**. These eyes look much like cut jewels; they are made up of many lenses set close together. Each lens sees a small part of what the ant is looking at. Together, the lenses form a fragmented picture of the whole. Compound eyes allow ants to see movement easily. Males and queens also have three simple eyes on the top of their heads called **ocelli**. These eyes distinguish between light and dark.

The mouthparts are made up of mandibles and maxillae. **Mandibles** are a pair of jaws that move from side to side. Ants use them for fighting, for digging, and for carrying objects. Smaller mouthparts, called **maxillae,** are found behind the mandibles and are used for chewing food. They have a row of tiny hairs that the ant uses to clean its legs and antennae.

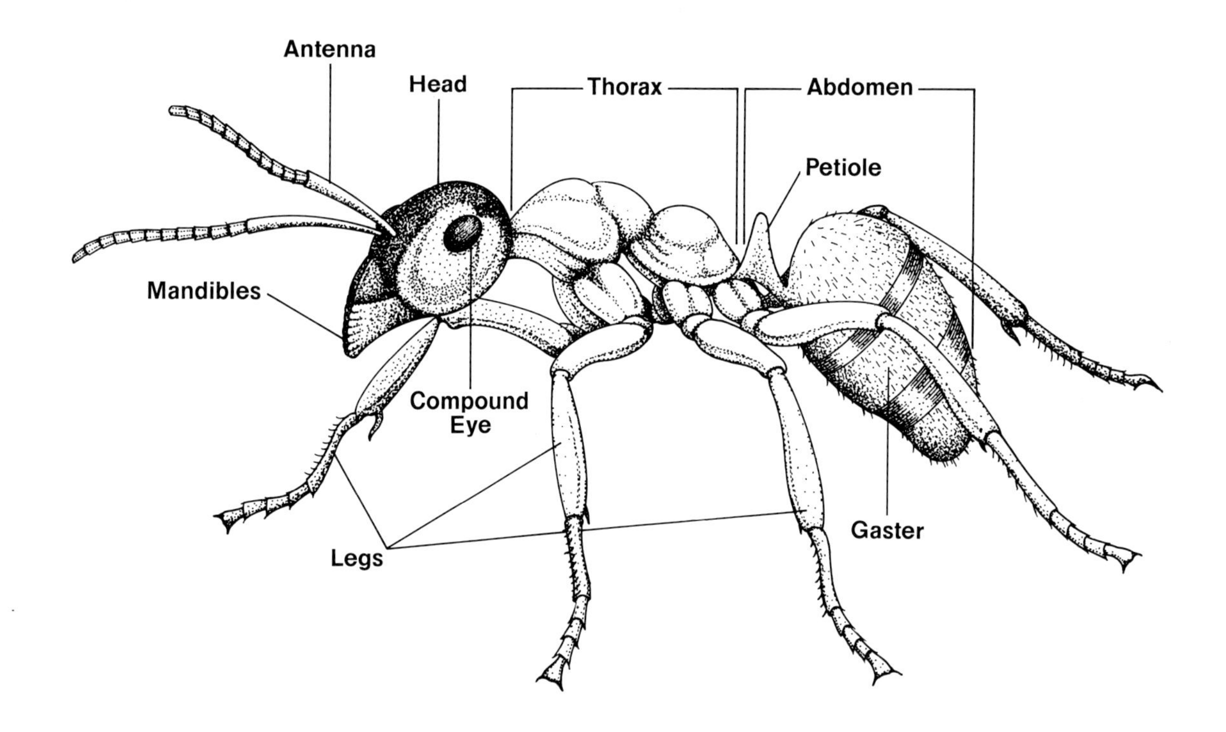

The thorax is the middle section of the ant. Three pairs of legs are attached to the ant's thorax. The foot of each leg has two hooked claws that make it possible for ants to walk upside down and to climb trees, as well as other vertical surfaces. Some ants use the claws on their front legs for tunneling underground. The front legs also have a row of tiny hairs much like those on the maxillae. And like those on the maxillae, these hairs are used to clean the antennae and the other legs. Queens that have not mated and males also have wings attached to their thoraxes.

The **abdomen** is made up of the **petiole**, or waist, and the **gaster**, or the enlarged part of the abdomen. The petiole connects the thorax to the gaster. Ants look so much like wasps that scientists believe ants evolved from wasplike ancestors more than 100 million years ago. The major difference between these insects is that the ant's petiole, unlike the wasp's, is made up of one or two movable segments that have humps on the top side.

An ant's gaster contains the crop and the intestine. Some ants, such as the red ant, also have a poison gland in their gasters.

When protecting their nests or killing other insects for food, red ants take their fighting position. They support themselves to the rear with their hind legs and lift their abdomen. This places them in a ready position to spray their poison, **formic acid**.

Formic acid is a substance that people have found to be very useful. At first it was used as an insecticide and an antibiotic. For a while, red ants were the only industrial source of formic acid. Now it is artificially made and can be used as a food preservative and as a disinfectant. Although contact with a few ants spraying formic acid isn't harmful to people, the amount of formic acid being sprayed by the thousands of ants in this picture is enough to asphyxiate a person.

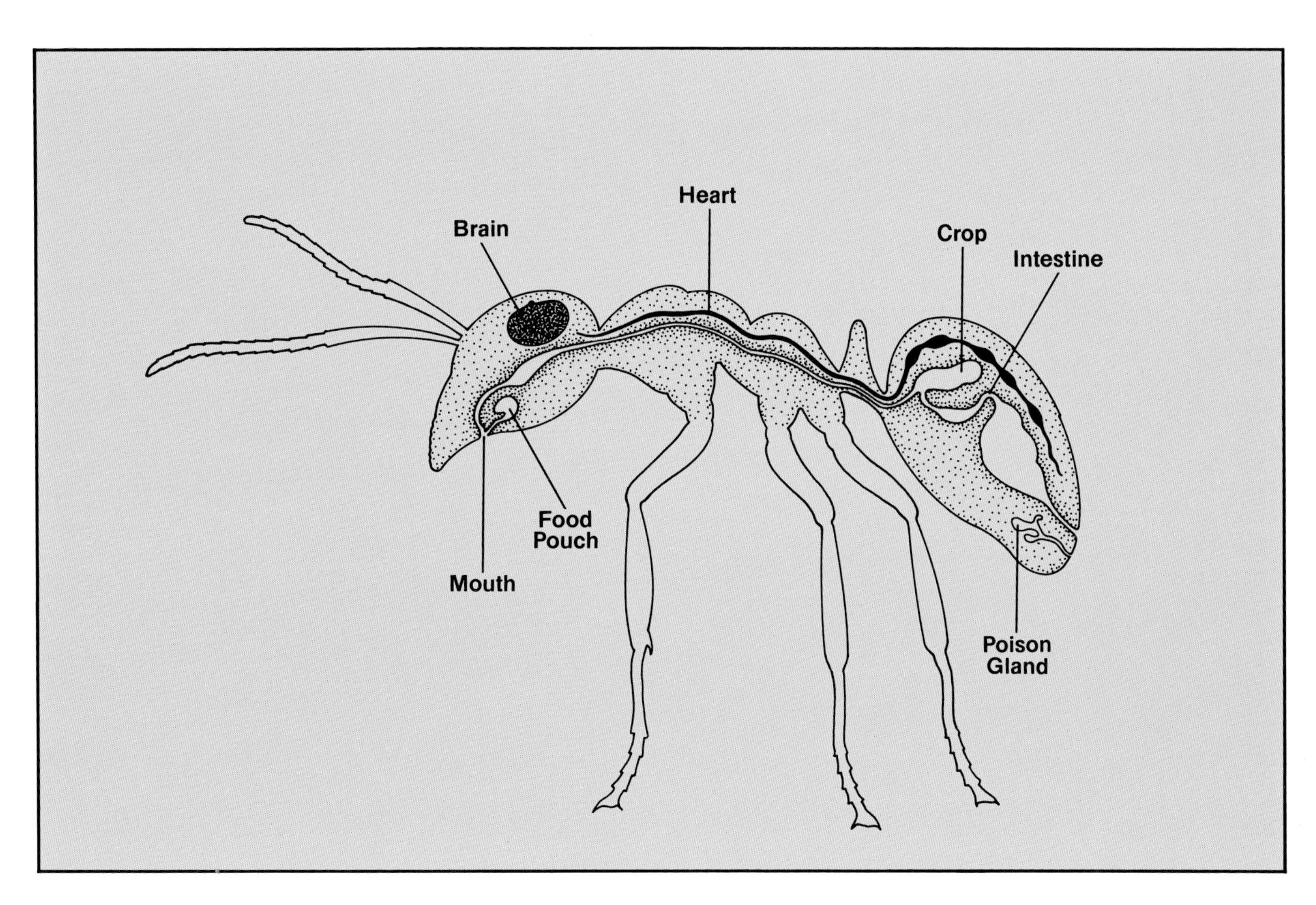

Ants eat by depositing chewed food in a pouch located below the mouth opening. The pouch muscles squeeze liquid out of the food. Ants swallow the liquid and spit out the solid parts of the food. The liquid is then stored in the **crop.** Whenever an ant with a full crop is hungry, some food from its crop travels into its intestine to be used by the ant's body. The crop is large enough that this does not significantly deplete its supply of food.

Since not every ant gathers food, the contents of the crop is also used to feed other ants in the nest. A hungry ant uses its antennae and legs to tap and stroke a food gatherer on the head. This special message signals the two ants to put their mouths together, and food passes from the food gatherer's crop to the hungry ant. An ant with a full crop can distribute food to 8 to 10 other ants. These ants, in turn, share their food with others. In this way, one ant can distribute food from its full crop to more than 80 other ants.

A favorite food of red ants is produced by aphids that suck juices from trees in the forest. These juices have more water and sugar than aphids need, so they excrete it in the form of a nutritious liquid sugar called honeydew. Ants tap and stroke the aphids with their antennae, signaling the aphids to release the honeydew. This process is called **milking**. The aphids also benefit from the ants. Red ants protect aphid colonies from other insect predators, such as ladybugs and their larvae, that like to eat aphids. Certain caterpillars that feed on the same trees that the aphids feed on—thus threatening the aphids' food supply—also make tasty meals for red ants.

The workers are now busy repairing and enlarging their anthill. They drag twigs, pine needles, and other plant debris to the nest. The weight of these building materials can be 10 to 20 times the body weight of an ant. One to two hundred ants together weigh only one gram. Any piece too large or too heavy for one ant to carry is handled by teamwork. Ants rarely give up, so with their united power they succeed in bringing building material or food back to the anthill.

Ants are very clean animals. The chambers within an anthill are constantly being cleaned and reordered by the busy workers to keep rot and mildew from occuring. When an ant dies inside the anthill, a worker immediately carries it outside to a distant cemetery.

Visible paths can be seen surrounding anthills. These paths may lead to the cemetery, to other friendly ant colonies, or to productive food sources. Special workers, called hunters, search for food within a radius of about 170 feet (50 m) from the anthill. Beetles and caterpillars that are much bigger than the ants, are attacked and numbed by the ants' poison spray. Ants not only attack and kill nonpoisonous insects, they also successfully attack bees and wasps. In one day, the hunters of a large colony can capture thousands of insects and drag them into the anthill.

As skilled as they are at destroying insects many times their own size, ants, too, have enemies. Woodpeckers use their long, sticky tongues to pierce the nest and eat ant larvae and pupae. Parasites in the form of mites live directly on the ants' bodies and slowly kill them. Other ants, even those of the same species, may be treated as enemies. But the red ants' worst enemies are people who purposely spray insecticides and destroy their nests.

Great excitement breaks out among the ants when an anthill is disturbed. The ants immediately begin to repair the damage. At first glance, the activity appears confused, but a careful observer recognizes that every ant does a specific job. Some carry the pupae away, others act as guards and secure the openings, still others find and bring new nesting material to repair the damage. Some ants just stay still. They are temporarily closing the nest entrances with their bodies so heat and humidity don't escape from the nest.

The year is coming to an end; it is now autumn. The larvae and pupae have matured. The ants become sluggish. They plug up entrances to their nest to protect the anthill from the coming cold. Then they retreat 20 to 80 inches (.5 to 2 m) into the anthill and remain together, motionless, within the tunnels and chambers. There they rest until the spring sun awakens them.

abdomen: the rear portion of an insect's body

antennae (singular *antenna*): a sense organ on the heads of insects

brood: the young of an animal, such as the egg, larva, and pupa of an ant

cocoon: a covering that a larva usually forms around itself in which it passes the pupal stage

colony: a group of social insects, such as ants, that live together and share work duties

compound eyes: eyes made up of many lenses

crop: an enlargement in the gaster that is used to store food

egg: an animal's female reproductive cell

exoskeleton: the hard, protective covering of an insect

formic acid: a liquid acid that some ants spray as a defense or as an aid in killing other insects

gaster: the enlarged part of the abdomen that, in ants, contains the crop and the intestine

larvae (singular *larva*): an immature form of insect that hatches from an egg

mandibles: a pair of jaws that move from side to side

maxillae (singular *maxilla*): small mouth-parts that are behind the mandibles

milk: the process ants go through to stimulate aphids to produce honeydew

ocelli (singular *ocellus*): small simple eyes found on the top of some insects' heads

petiole: the waist, or segment, that joins the rest of the abdomen to the thorax in some insects

pupae (singular *pupa*): the stage when larvae, often enclosed in a cocoon, change into the adult form

pupate: to pass through the pupal stage

queen: a fertile (able to reproduce) female social insect, whose primary job is to lay eggs

sperm: an animal's male reproductive cell

thorax: the chest, or midsection, of an insect

worker: a female social insect that performs most of the jobs for the colony

INDEX

abdomen, 24, 28
antennae, 27; and communication, 33, 35
anthill. *See* nest
aphids, 35

bees, 7, 39
beetles, 14, 39
body of ant, 24, 27, 28; diagram of, 27, 28, 31; size of, 24, 36
brood, 8, 9, 17

caterpillars, 35, 39
cocoon, 21; hatching from, 22
colony, 7, 8, 12, 39
communication, 27, 33, 35
compound eyes, 27
crop, 28, 31, 33

eggs, 8, 12, 14-15, 17, 21
enemies, 41
exoskeleton, 24
eyes, 27

fighting, 27, 29
flight. *See* mating flight
food, 12, 19; and eating, 27, 31; gathering of, 14, 29, 33, 35, 39; sharing of, 33
formic acid (poison spray), 29-30, 39
Formica polyctena (species name), 5

gaster, 28

head, 24, 27
honeydew, 35
hygiene of ant, 17, 27, 28, 36
hunter, 39

insecticides, 30, 41
intestine, 28, 31

ladybugs, 35
larvae, 8, 14, 19-21, 45
life span of ant, 8

male ant, 8, 11-12, 27, 28
mandibles, 27
mating flight, 11-12
maxillae, 27, 28
milking, 5, 35
mites, 41
mouthparts, 27

nest, building of, 9, 12, 14-15, 36, 43; diagram of, 14-15; disturbance of, 41, 43; protection of, 29, 43, 45; size of, 9; temperature of, 9-10, 17, 19, 21, 43; nurseries, 15, 17
nursery, 15, 17
nurses. *See* worker

ocelli, 27

parasites, 41
paths made by ants, 39
people and ants, 30, 41
petiole, 28
poison: gland, 28; spray, 29-30, 39
pupae, 8, 14, 19-21, 45
pupate, 19-21

queen, 8, 11-12, 14, 15, 27

range of ants, 5

saliva, 17, 19
size of ant, 24, 36
social insect, 7
sperm, 11, 15
strength of ants, 5, 36

thorax, 22, 24, 28
types of ant, 8

wasp, 28, 39
woodpeckers, 41
worker, 8, 10, 12, 21, 36; role in nest building, 14, 36; role as nurses, 15, 17, 19, 22; role as food gatherers, 14, 19, 33, 35, 39

ABOUT THE AUTHORS

Heiderose and Andreas Fischer-Nagel received degrees in biology from the University of Berlin. Their special interests include animal behavior, wildlife protection, and environmental control. The Fischer-Nagels have collaborated on several internationally successful science books for children. They attribute the success of their books to their "love of children and of our threatened environment" and believe that "children learning to respect nature today are tomorrow's protectors of nature."

The Fischer-Nagels live in Germany with their daughters, Tamarica and Cosmea Désirée.